# Sound and Music

## Kay Davies
## and
## Wendy Oldfield

Wayland

# Starting Science

Books in the series

Animals
Electricity and Magnetism
Floating and Sinking
Food

Light
Sound and Music
Waste
Weather

# About this book

*Sound and Music* investigates, in relation to children's early experiences, how sound is produced, transmitted and received. The themes develop children's aural skills through sound recognition, dynamics and pitch variation. They learn that sounds can be produced naturally and mechanically, and also how sound can make music and be useful in other ways.

This book provides an introduction to methods in scientific enquiry and recording. The activities and investigations are designed to be straightforward but fun, and flexible according to the abilities of the children.

The main picture and its commentary may be taken as an introduction to the topic or as a focal point for further discussion. Each chapter can form a basis for extended topic work.

Teachers will find that in using this book, they are reinforcing the other core subjects of language and mathematics. Through its topic approach *Sound and Music* covers aspects of the National Science Curriculum for key stage 1 (levels 1 to 3), for the following Attainment Targets: Exploration of science (AT 1), Types and uses of materials (AT 6), The scientific aspects of information technology and microelectronics (AT 12), and Sound and music (AT 14).

First published in 1990 by
Wayland (Publishers) Ltd
61 Western Road, Hove
East Sussex, BN3 1JD, England

© Copyright 1990 Wayland (Publishers) Ltd

Typeset by Nicola Taylor, Wayland
Printed in Italy by
   Rotolito Lombarda S.p.A., Milan
Bound in Belgium by Casterman S.A.

**British Library Cataloguing in Publication Data**
Davies, Kay, *1946–*
   Sound and Music.
   1. Sound
   I. Title  II. Oldfield, Wendy  III. Series
   534

ISBN 1 85210 998 X

Editor: Cally Chambers

# CONTENTS

The drummers play the drums with their hands.
The rhythm makes us want to dance.

# BEAT THE DRUM

You can use an empty biscuit tin to make your own drum.

Stretch a piece of cling film over the tin like a tight skin. Make sure that the edges are sealed.

Tap out a rhythm on your drum with your fingers or a stick.

Put some buttons and beads on the cling film.

Now tap your drum and watch them jump.

The skin shakes the air too, just like it shakes the buttons and beads.

All the sounds you hear are made when the air shakes or vibrates.

# WHAT'S THAT SOUND?

We hear sounds with our ears. We can learn to tell one sound from another.

Find some objects like these and play a sound game with your friends.

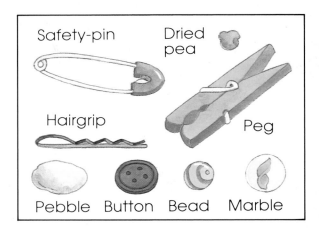

Put each object, one at a time, into a box. Shake it and listen carefully to the sound. Now take it in turns to put an object into the box secretly.

Shake the box while the others listen hard. See if they can guess what is making the sound inside.

The market is full of busy people buying their food.
There are noises all around.

The rattlesnake shakes its tail. The rattling noise warns people and animals to keep their distance.

# SHAKE AND RATTLE

We can make rattling sounds like the rattlesnake.

Put some pebbles in a tin. Shake your tin and listen to the loud sounds.

Put some sand in another tin. Shake your tin and listen to the soft sounds.

Try objects like marbles, buttons, rice, beans, sugar, crayons and paper clips in your tin. Test which make loud sounds and which make soft sounds.

Make different rattles with your objects.

Try using a glass jar, a plastic tub, a paper bag and a cardboard box.

Do they all sound the same? Can you try to describe the sounds?

9

# POT AND PAN BAND

A pot and pan band can make lots of sounds.

Make a collection of pans, lids, spoons, mugs and other kitchen things.

Tie them with string and hang them from a pole.

Rest your pole firmly on the backs of two tall chairs.

Use drumsticks or a wooden spoon to tap all the things. Listen for high sounds and low sounds.
Try to put them in order from low to high.

Hit your pots and pans, then hold them to stop them shaking. What happens to the sounds now?

Music played on the steel drums is fun to listen to.
The drums make high sounds and low sounds.

# A GOOD BLOW

Whistles make sounds when we blow into them.

Find a straw to make your own whistle.

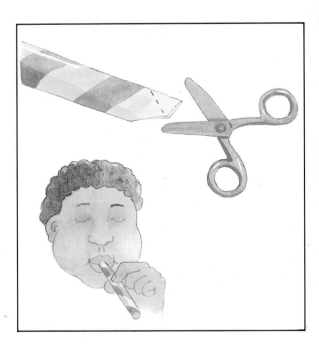

Flatten one end and shape it to a point with scissors.

Practise blowing into the pointed end to make a loud sound.

Cut some straws to make whistles of different lengths.

You can blow high sounds with short straws and low sounds with long straws.

Glue your whistles together in order of length to make pan-pipes like these.

The referee blows the whistle loudly so that everyone can hear. They stop playing the game.

The harp is a musical instrument with lots of strings.
We pluck the strings with our fingers to play tunes.

# LOTS OF PLUCK

String, wire or elastic can be pulled tight and plucked with fingers.

The wire strings on this guitar vibrate and make music.

Plucking long strings makes low notes.

Plucking short strings makes high notes.

You can make your own guitar from an old shoebox.

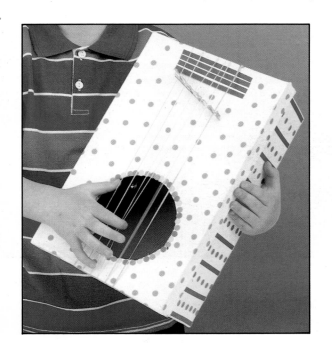

Cut a circle out of one end of the lid.

Use this to cut a bridge for your strings. Fix it at a slant on the lid.

Stretch large rubber bands around the box and lid. Make sure they rest on the bridge.

The waves crash on to the beach. The water rattles the stones and shells as it goes back out.

# SOUNDS THE SAME

You can copy the sound of waves on a beach.

Put some small stones into a cardboard box. Fill a jar with water and put a straw in it.

Ask a friend to gently roll the box while you blow down the straw.

You can make up a play and use sound effects.

A pencil rubbed over a comb sounds like winding a clock or toy. Blocks of wood tapped together sound like footsteps. Blowing across the open top of an empty bottle makes a sound like a ship's hooter.

# SOUND AN ALARM

Bells, sirens and whistles can all be used as alarm and message sounds.

They usually make high, loud sounds. They can be heard clearly from a long way off.

Here are some different alarm and message sounds. Are they bells, sirens or whistles?

Answer the telephone

Open the door

Wake up!

Late for school

The kettle is boiling

Get out of the way

Make a chart of other message sounds. Tick off all the ones you have heard. Try adding to your list.

| | Bell | Siren | Whistle |
|---|---|---|---|
| | | * | |
| | * | | |

The siren on the ambulance gives a warning. It tells us to clear the road and take care.

# A LOUD VOICE

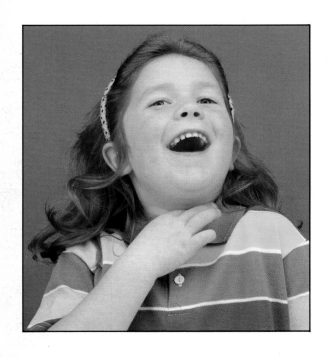

We can feel sounds.

Gently rest your fingers on your throat and talk to a friend.

Then try whispering, shouting and laughing.

A loud voice makes your throat vibrate more than a soft one.

Speaking through a funnel makes a voice very loud.

Make this megaphone with card and sticky tape. Be sure it is open at both ends.

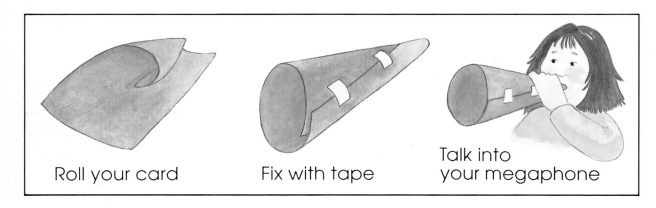

Roll your card          Fix with tape          Talk into
                                               your megaphone

Speak to your friend with your megaphone and then without. Ask your friend which sounds loudest.

The man uses the megaphone to talk to the crowd,
so everybody knows what is going on.

The jack rabbit's big ears collect sounds. It can hear noises from all around.

# BIG EARS

If you hold a megaphone to your ear it picks up sound. It becomes an ear trumpet.

Ask a friend to whisper a message so that you can just hear it.

Then use your ear trumpet and see how much louder the message becomes.

Do the test again. Turn your ear trumpet away from your friend's voice. Turn it towards the voice again. Is there any difference?

Look at the shapes of these animals' ears.

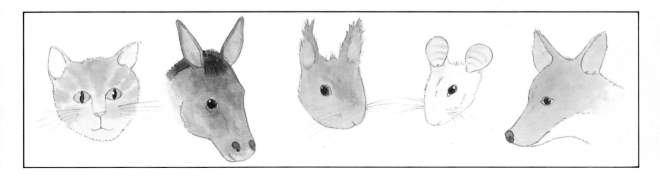

They can move their ears to pick up sound from behind and in front.

# CAN YOU HEAR ME?

You can make your own telephone.

Find two plastic pots and a very long piece of string.

Thread the string through a hole in each pot. Tie it with knots.

Ask a friend to hold one pot to their ear.

Pull the string tight between you.

Speak into your pot and your friend will hear you quite clearly.

Can you make a better telephone?

Try different lengths of string, nylon thread and wire. Try using tins or paper cups.

You could cover each open end with a piece of balloon or cling film, and fix it with a rubber band.

We use the telephone to talk and listen. We can speak to our friends in other towns and countries.

The machine makes a nasty noise. The woman wears
ear mufflers to protect her ears.

# WHAT A DIN!

Some sounds, like machinery or loud music can hurt our ears.

We need to stop really loud sounds reaching our ears.

Cover your ears with your hands. Sounds become quieter and not so clear.

We can make places quieter too.

Put a ticking clock into a tin box. Then put the clock into a cardboard box.

Which sounds louder?

Now put some newspaper into the tin box with the clock. Replace the lid and listen. Can you stop all the sound?

# SILENCE

When all is quiet, sit very still and listen hard.
Is there silence or can you hear faraway sounds?

Machines make noises. You may hear cars, a lawn mower, a drill or a television.
People and animals make noises. You might hear a bird singing, a dog barking or someone talking.
Nature makes sounds too. The wind whistles and the rain splashes.

Draw up a chart of all the noises you hear.

Machines          Living things          Other sounds

Our world is never really silent.

Even in the quiet of night, your bed springs might creak and break the silence.

The moon has no air. There are no sounds on the moon.

# GLOSSARY

**Bridge** A support for the strings on a musical instrument.

**Ear mufflers** Pads worn over ears to quieten sounds reaching them.

**Ear trumpet** A funnel-shaped instrument which collects sound, making it louder.

**Megaphone** A funnel-shaped instrument which throws out the sound of a voice, making it louder.

**Music** Lots of notes which are played or sung together.

**Note** A musical sound.

**Pluck** To pull and then let go of a string, making it vibrate.

**Referee** A person who makes sure that players keep to the rules of the game.

**Rhythm** The pattern of beats in music.

**Siren** An instrument or machine that makes a loud wailing sound.

**Sound effects** Sounds made to copy the real thing.

**Vibrate** To shake up and down quickly.

**Wire** A long piece of thin metal.

# FINDING OUT MORE

**Books to read:**

**Hearing** by Wayne Jackman (Wayland, 1989)
**Hearing** by Henry Pluckrose (Franklin Watts, 1988)
**Hearing** by Anne Smith (Firefly, 1989)
**Musical Instruments** by Jenny Wood (Franklin Watts, 1989)
**Musical Instruments** by Alan Blackwood (Wayland, 1987)
**My Class Makes Music** by Vicki Lee and Daniel Thomas (Franklin Watts, 1986)
**Scrape, Rattle and Blow** by Chris Deshpande (A & C Black, 1988)
**Sound** by Angela Webb (Franklin Watts, 1987)

The following series may also be useful:

**Noises** by Wayne Jackman (Firefly Books, 1990)

## PICTURE ACKNOWLEDGEMENTS

Cephas Picture Library 11; Chapel Studios 15 top, 26; Eye Ubiquitous 13, 16, 21, 25; Frank Lane Picture Agency 22; Hutchison 4, 7, 12, 14; J. Allan Cash Ltd. 19; Survival Anglia Ltd. 8; Wayland Picture Library (Zul Mukhida) cover, 5 both, 6, 9 both, 10, 15 (bottom), 17 both, 20, 23, 24, 27 both; ZEFA 29. Artwork illustrations by Rebecca Archer.
The publishers would also like to thank Davigdor Infants School, St. Andrews C.E. School and Somerhill Road County Primary School, all of Hove, East Sussex, for their kind cooperation.

# INDEX

Page numbers in **bold** indicate subjects shown in pictures, but not mentioned in the text on those pages.